To Charlie Kane
Director of the Orleff Gallery

All rights reserved. No part of this book may be reproduced or transmitted in any form or by any means, electronic or mechanical, including photocopying, recording, or by any information storage and retrieval system, without permission in writing from the author or Orleff Galleries.

Please address all inquiries to Orleff Galleries: fax: 973-325-0930 or email: orleffgallery@gmail.com (for English).

Paperback books may be purchased online through Amazon.com and other retailers. Hardcover copies may be purchased directly from the Orleff Galleries by contacting us via email: orleffgallery@gmail.com.

Copyright © 2014 by Orleff Galleries.

Published by Orleff Galleries

ISBN-10: 1495988643

ISBN-13: 9781495988646

Library of Congress Control Number:

The Proof of Inherited Acquired Characteristics

The Death of Random Mutations

Yuriy Kalinnikov
(Posthumously)

Contents

AUTHOR'S PREFACE VI

CHAPTER I 1
BACKGROUND 1
   INTRODUCTION 1
   HISTORY 6
   SCIENTISM -MATERIALISM 6
   EVOLUTION 11
   DARWINISM 11
   NEO-DARWINISM AND GENETICS 12
   1900 AND BEYOND 13

CHAPTER II 15
EVOLUTION 15
   LAMARCKISM 15

CHAPTER III 22
NATURAL SELECTION – GENETICS AND
RANDOM MUTATIONS 22
   NATURAL SELECTION 22
   GENETICS 28
   RANDOM GENETIC MUTATIONS 30

CHAPTER IV 32
RECENT GENETICS 32
   EPIGENETICS 32

CHAPTER V 39
ENERGY – MATTER – MIND 39
   QUANTUM MECHANICS (PHYSICS) 39
   UNPREDICTABILITY 40

CHAPTER VI
MATERIAL BIOLOGICAL SCIENCES VERSUS NONMATERIALISM 47
THE BATTLE LINES 47

CHAPTER VII 57
INHERITED ACQUIRED CHARACTERISTICS 57
THE EXPERIMENTS 57
DEGENERATIVE AND ATROPHIC CHANGES 59

CHAPTER VIII 66
THE ESTABLISHMENT VERSUS KAMMERER 66
THE ATTACK 66

CHAPTER IX 80
DISCUSSION 80
INHERITED ACQUIRED CHARACTERISTICS 80

CHAPTER X 87
THE DEEPER MEANING 87
IMPLICATIONS 87

CHAPTER XI 91
KALINNIKOV'S HOLISTIC THEORY 91
HOLISTICS 91
MIND 92
EPIGENETICS 94
GENETICS – INTRONS AND EXONS 95

CHAPTER XII 100
FINAL PROOF OF ACQUIRED CHARACTERISTICS 100
END 106

PREFACE

This book is a composite of notes that Kalinnikov gave me to read, possibly never intending to have them published to which was added a brief history of the development of inherited acquired characteristics, summaries of his theory, details about the Kammerer incident leading to his death, and finally proof of inherited acquired characteristics.

Certainly, the attacks by the scientific community on Lamarck initially and then on Kammerer and inherited acquired characteristics were undeserving along with the journal, *Nature* printing opinions without demanding supportive evidence. There have been other times when the established community strongly resisted new discoveries such as epigenetics. Quantum physics was often described as fantasy in the mind of Neils Bohr, and discouraged trying to have Kalinnikov's books see the light of day. The field of paleoanthropology often perpetuates disproven concepts and negative conjectures.

However, these criticisms should not overshadow the majority of open-minded scientists in all fields who remain receptive to new ideas and developments and regret the need to include the science or scientists with strong adverse reactions to new discoveries and interpretation.

This book has not been edited but it speaks for itself in the raw language of the writer which hopefully does not distract from the issues covered. Despite its coverage of old issues, it does put to rest the idea that random mutation which somehow produces new mutations willy-nilly in organism that are then "selected" by natural selection, instead of the fact that organism can change themselves in a directed and more timely fashion. The new proof of inherited acquired characteristics destroys the third and last support that natural selection has rested upon and with Darwin himself abandoning it, what can save it now?

Despite it acerbic style, we trust the reader will find that information enlightening and open the door to new ways of looking at evolution.

Acknowledgements

As in the past our good friend Charlie Kane has made possible the printing of this and previous books. I speak for Yuriy and myself in expressing our deepest appreciation for all he has done.

I. Kalinskiy, Moscow

The Proof of Inherited Acquired Characteristics

The Death of Random Mutations

Chapter I

Background

Introduction

On Sept 23, 1926, Paul Kammerer was found dead in the woods near Vienna Austria, shot above the *left* ear and the gun in his right *hand*. Thus began one of the great unsolved mysteries of our times. While suicide was intended, why was the gun in the wrong hand? And if he shot himself on the left side, why wouldn't the gun drop to the ground on the left side?

Paul Kammerer (1880-1926) was the most famous biologist of his day. He championed and revised interest in inherited acquired characteristics in the 1920s. He was becoming famous world-wide. The NY Times stated that he was the successor to Darwin. Einstein was quoted as saying: (Kammerer's) work is original and by no means absurd. Russia had always accepted Lamarck and acquired characteristics since it was first written, and Kammerer was well respected even in the Soviet Union. He was invited by none other than the famous Ivan Pavlov to a professorship at the University in St. Petersburg (then, Leningrad) before his untimely demise.

The "suicide" was attributed by the press to "shame due to fraud" because it was discovered that one of his specimens had been fraudulently injected with India ink to make it look like the real nuptial pad of the toad in the experiment years after many experts in England have examined it and found to be authentic. The director of the Institute along with Kammerer felt that someone had purposely done the deed much later to discredit Kammerer, some days earlier when they knew an expert from the United States was going to examine it. It had to be done after the specimen was studied in England and shortly before G. K. Noble was to examine it. Otherwise the India ink would have spread out. It was doubtful if Kammerer even had access to the specimen. However, because of other circumstances in his life in Vienna at the end of WW-I when the former capitol of the Austro-Hungarian Empire had been totally defeated and bankrupt along with Kammerer who lost everything. He had reasons to take his life, but his suicide easily convinced the press that shame due to fraud was the reason.

Despite the accolades from many scientists and the press earlier, he had many enemies. The supporters of the purely materialistically natural selection and random genetic mutations theory advocates saw him as enemy number one. With Kammerer out of the way and disgraced the field of study of inherited acquired characteristics was silenced, especially because of the brutal attacks and innuendos and outright distortions of the truth that had led to the condemnation and accusations of dishonesty in the years prior to his death. The reigning theories of evolution (natural selection and random genetic mutations) remain the official theories today despite overwhelming evidence against them. Acquired characteristics had been almost forgotten again.

However, the issue was not over. A writer Arthur Koestler in 1971 investigated the entire incident. He was originally convinced of Kammerer's guilt when he started his investigation, but ended up feeling that Kammerer was not involved in the fraud and that it was a conspiracy to defame Dr. Kammerer. It is one of the most completely investigated cases in science and could only be written by one dedicated to science and the perseverance to uncover all the facts, publications and correspondence surrounding the case. The book, the *Case of the Midwfe Toad,* resulted not only in exonerating Dr

Kammerer, but exposed the viciousness of the scientific community in discrediting an opposing theory. Now that inherited acquired characteristics have been proven, the story is even more poignant.

However, recently, new evidence and the new sciences as epigenetics has put acquired characteristics back in the forefront of evolution and by other new discoveries and have proven that inherited acquired characteristics to be the correct interpretation. These new sciences; were only established in the last hundred years (paleontology and genetics) or fifty years (epigenetics), and even more recently in the twenty-first century (molecular evolution).

Still, the biological scientific community has such a large stake in the older natural selection and random mutation theories that they resist acknowledging the recent findings. The older accepted theories reign supreme. Actually, the stakes as to which theory is the accepted one are enormous. To think that this was merely a scholarly debate between opposing theories would be naïve.

In the world of science, entire reputations could be destroyed depending on the outcome. Financial support depended on which is the accepted theory and the direction of research. The accepted theory receives the professorships, grants, and press coverage. They alone become the authorities as society goes along with their convictions. Even in science today, the system works this way and the old guard is reluctant to cede credence to the new. A famous example is the new science of quantum physics, which rocked the old classical materialistic physics. Einstein, a leader of the old guard tried to disprove it for the last forty years of his life and failed and the new generation of physicists joined the ranks of the new quantum physics. Even the public has not yet grasped the implications of quantum physics, which has far wider implications than a new theory of evolution, and adds essential factors that give evolution an entirely new direction and understanding.

Because, today the old theories are deeply ingrained in all aspects of the biological sciences—entire books, the teaching of evolution and general mindset would have to change to accept the new proofs—it is far from easy. Also in most present day biological sciences, natural selection and random genetic mutations have *already* been established as facts, and no longer the active subject of research. Yet, less than half of the public has accepted it, as opposed to those religiously inclined who still accept Genesis. Many find these theories unconvincing.

History

Scientism -Materialism

The ruling conviction since the 17th century is that energy and matter are the only aspects that compose the universe that one could see, touch, or feel or perceived through scientific instruments. The conviction was formulated in the seventeenth century by Rene Descartes (1596-1650) who claimed that the mind was spiritual and the body was mechanical (material) and were separate and only communicated through God. The mind and soul belonged to God and the body (material)) belonged to science or so Descartes defined it. It became the new dogma of science, even today. Everything *had to have* a material explanation. The universe, matter and energy are all happenstance with no guiding hand. The laws of science were developed to be as independent of any supernatural or religious beliefs as possible, and any metaphysical effects (changed to mean supernatural) were against all rational thought. This view is often called Scientism or Materialism.

These are the reigning theories of science, ever since.

Aristotle coined "metaphysics" as anything outside science such as art, philosophy, literatures, etc. which these scientists (called "scientismists") have given the word "metaphysics" a pejorative connotation—supernatural. These same scientists have the belief that these other disciplines cannot describe the universe and are undeserving of attention and should even be discouraged and ignored. This belief in the redundancy of philosophy and other metaphysical explorations is now supposed to be unnecessary as scientism could explain any phenomena and solve all economic problems as well. In some circles it is more discretely defined as the only true way to explain reality. It has become very current in the halls of academe and in a sister discipline— socialism (defined as a materialistic Hegelian dialectic by the materialistic industrialist Friedrich Engels, a close associate of Karl Marx), a discipline founded on scientism. Materialism is all the rage in the West. Despite established quantum mechanics, and epigenetics, the pervasive interpretations are still based on materialism. Today, science is still the science of materialism.

It is the conviction of most scientists and the news media that matter and energy are the only things that exist in the universe. The materialists have convinced most non-deist people that their theories are correct. The public had no other information to question it, except of course the Bible for those who rejected materialism totally or tried to fit scientism into their doctrine. The theory had become an entire self-indoctrinating mindset. The way materialists *define* the problem, they can answer it, except awareness and consciousness which were left alone as "spiritual."

The problem with Lamarckism was that it had introduced a nonmaterial element in his theory which could not be explained by the materialists and sounded supernatural or deistic. Lamarck's acquired characteristics called this motivation "sentiments interieur" or urges and strivings, they would have to come from the mind that had been the province of God according to Descartes, and was not scientific. Therefore, the materialists had a need to negate nonmaterial theories such as inherited acquired characteristic of Lamarck, to keep the faithful (and themselves) convinced that their theory (and materialism) was flawless in all areas. However, this excessive zeal bespeaks closed minds to debate. But new facts won't go away, and older ones have been recovered (here, from Kammerer's own book!).

The theory of scientism was a rebellion against religious beliefs and the supernatural. However, all rebellions tend to be over simplistic and opinionated without looking at all the facts. They need society to endorse them with limited information and simplistic solutions. Because, it appears to be convincing (ignoring nonmaterial aspects of life), it has governed thought, philosophy, and science for the past two hundred years. There were no real challenging theories to contest it except religious thought in the opposing corner and the so-called invalid inherited acquired characteristics which had been dismissed as deistic. Materialists controlled science during this time until the present.

The real contention was not around chemistry and the laws of physics (yet!), but the split was between living things and the so-called material world. To explain life, they claimed it was purely materialistic and all species below humans were biochemical automatons, put together by chance combinations of RNA or DNA and other organic compounds. Only humans (from Eden?) could think and it should eventually be explained as some higher materialistic biochemical process involving neurons, but nothing so far has seemed to be convincing. Because of later animal research and the acceptance of the fact that animals have minds, some of this thinking has gone by the wayside—quietly.

Natural selection theory started out as general conviction of materialistic scientists that spread throughout society as the official belief, based on the underlying theory that only matter and energy existed (materialism) and anything else is supernatural. Once such a belief (conviction) settles in, as opposed to a religious interpretation, the new dogma becomes the new conviction of the day, especially for those who reject the old religious formulations. About 39 percent of the population slowly adopted it as the more "modern" interpretation. All things become influenced with this new interpretation, science, literature, and philosophy, along with it.

Entire generations are imbued with this belief or theory. To unseat it or change the meaning changes a great deal of society and their philosophical outlook. This can be seen in the change from the transition from religious beliefs to the materials science that is still battling on.

Evolution

Lamarck's theory (1809) was the first theory of evolution but had introduced *nonmaterial* factors as urges and strivings before any materialist formulated a theory of evolution. At the time, Lamarck's theory was distorted and ignored in the West. They were looking for a materialistic theory to replace Genesis and any other theory as Lamarck which was not totally based on materialism was rejected.

Darwinism

In 1859 Darwin published *On the Origin of Species*. It was a materialistic theory and grabbed by the materialists as the answer to a complete understanding of the world as a completely materialistic existence (despite naysayers in the background). The basis is the fact that individual traits vary. If an extreme trait fits better in the environment, it would prosper and struggle to become the fittest. **Even Darwin gave up on this theory** (as he confided in a letter to his cousin Galton) and changed to Lamarck's inherited acquired characteristics, which is noted in the 6th edition of his book which Darwinists rarely read. Still, the original Darwin is preached as gospel.

Natural selection filled the final gap in materialism by including the biological sciences.

To the average layperson, this may sound incredible that science is not founded on facts but on theories that are no longer valid, especially in evolution but society is uninformed of the real situation. Darwinism lives on as the standard theory. However, Darwinism as a theory was being abandoned in the late 1880s by many leaders in the field and coming to the same conclusion as Darwin himself—inadequate to explain evolution.

Neo-Darwinism and Genetics

Then, the later discovery of the work of Gregor Mendel on pea plants miraculously saved it. William Bateson coined the word "genetics" and W. Johannsen coined the word "gene." Neo-Darwinism was born based on the fixed traits of organisms. However the new concept drew theorists who interpreted Mendel far beyond Mendel's work on pea plants.

But to keep the status quo is destroying the credibility of science for the public who support the material sciences. The neo-Darwinists stated that all genes were fixed and no new traits were allowed.

By 1924, a chief founder and proponent of neo Darwinism, Bateson confided to his son that it was not working as a theory of evolution and regretted spending the last years of his life promoting it.

Before the 21st century there were no concrete proof of *any* evolutionary theory, only hypothesis, supposition, or conjectures existed. Experiments were possible with Lamarck's inherited acquired characteristic but natural selection could only be conjectured since it was supposed to occur over millions of years according to Darwin. If the theory was materialistic and logical, it became authoritative and observations were fit into it and others observation were ignored such as incipient form where no such traits or form existed previously (lungs in amphibians).

1900 and Beyond

Again Darwinism as neo-Darwinism was dying. The geneticists came to the rescue with a new idea based on the spontaneous rate of gene production. Genes reproduce every generation and some become mutated. The rate was considered one mutation in every 500,000 reproductions, but can vary with radiation and other factors. They reasoned that these myriads of mutations (especially in large herd reproductions) would be available for natural selection to choose from. It sounds good but in small hominid population where numbers and rate of reproduction was small, it would be extremely limited. Today, the public only hears the dogma of random mutationists that now keep natural selection afloat.

Natural selection and random mutation theories had a backing of genetics, but with serious flaws that were glossed over. Genetic variations just happened on a regular basis it was theorized but whether the mutation was advantageous, neutral or cause serious problems was completely happenstance. One major flaw (among many) was incipient forms—that is, when a new structure or function evolved where no variation of traits was present. It is still ignored.

Having destroyed Paul Kammerer and the advocacy of inherited acquired characteristics, there was no competition and natural selections and random mutations rule supreme until today in every science book mentioning evolution.

Chapter II

Evolution

Lamarckism

Then in 1809, a brilliant Frenchman and noted scientist Jean Baptiste Lamarck came up with the *first* scientific theory of evolution. It is based on inherited acquired characteristics. He was the first to break away from the Bible and formulate a theory based on natural laws of science.

He saw all animals in terms of human beings who had needs, urges, and strivings and who looked for food and protected themselves and avoided danger to survive—just as we are motivated by urges and strivings. He justifiably assumed that all organisms are also motivated in the same way, making it an all inclusive theory of evolution of all life including humans as well. This implied that the organism or animals had *nonmaterial* urges and strivings and that they themselves were always on the lookout for new and better food sources, and protection to survive and not just mechanical beings. By strengthened various parts of the body in striving for new and more abundant food sources or to protect themselves, they increased their capacities in that direction (non-randomly) and was passed on to subsequent generations. It agreed with folk wisdom.

This implied that all organisms, including plants and animals, have feelings and are alive and have *nonmaterial* urges and strivings like us. It was very rational. However, it implied some metaphysical functions (supernatural) which disagreed with pure materialism, whether it was rational or not. Subjective tendencies according to materialists were deistic or supernatural as defined by Descartes. Until some totally materialistic theory of evolution appeared, there was no real controversy. Lamarck was considered interesting but that was about all. The search was on for a purely materialistic theory.

Lamarck was the first to break with Genesis and stimulated further research in evolution. He is truly the father of evolution.

True, Lamarck had made several assumptions in the theory that were later proven false (just as Darwin had) but which were blown out of proportion but not his basic assumptions regarding acquired characteristics. For example, he felt that life came from spontaneous generation in the ground and was a continuous process, thus we see various species at various levels of evolution rather than species that had plateaued at various levels. The other main objection was his belief that evolution was evolving to perfection. The materialists grabbed this statement to mean evolving toward God.

After the materialistic theory of natural selection was accepted, Lamarck's theory was discredited at every chance the materialists got, even to the point of aggressively attacking what they termed deistic such as inherited acquired characteristics a *nonmaterial* deistic phenomenon.

Lamarck's theory violates the pure materialistic interpretation of materialistic theory where no such nonmaterial ("metaphysical") element should exist. Things happen by pure physical laws free of any metaphysical (translate:-- supernatural) properties—pure random phenomena—chance, nothing more. Therefore, materialists don't need a supernatural deity to explain anything (nor most of us for other reasons). Therefore, Lamarck's theory of evolution was totally rejected especially in England and by Darwin in particular (at first). To further humiliate Lamarck, the English mistranslated Lamarck's interior sentiments as "intension" to make it sound ridiculous. Any first year French student could have translated it correctly as "internal sentiments" or urges. It became vogue to negate acquired characteristic in the West with ludicrous accusations, while here in Russia; it was the accepted theory from the start and agreed with the prevailing folk wisdom, even under communism where Kammerer was praised by the soviets until about 1940 when it was supplanted by natural selection and random genetic mutations. Earlier, Kammerer had been invited by none other than Ivan Pavlov to be a professor at the State University before his demise.

The hyper-criticism amounted to distortions of Lamarck's theory and drew irrational conclusions to preserve the established dogma of materialistic scientism.

On the other hand to do everything to debase the opposition is not uncommon. When a British zoologist published a paper entitled *What Lamarck really Said* during the centennial of Darwin's book, pointing out some of the grosser errors and misstatements about Lamarck, they intensified their same attacks with the same aggressive inaccurate criticism that were ever perpetuated against Lamarck.

The English totally influenced American thinking against Lamarck in favor of natural selection. All Western books on evolution repeat these absurd accusations. In the West, it was totally discredited implying that Lamarck said that animals change by conscious *intention*, made the theory sound ridiculous—even laughable. Lamarck never said it. However, it was survival at stake in the science world and no holds were barred. In America, they bought the description without any attempt to read Lamarck's original writings.

But had any one bothered to notice, his theory is based on far more rational considerations that he derived from a background that transitioned through geology, paleontology (not yet a science but from his own work in taxonomy), botany and zoology. Because of his wide experience as a taxonomist, it positioned him well to write a theory of evolution based on natural science that broke completely with the Bible. In France the church had been overthrown during the reign of terror and he was free of such restraints. He had a much greater depth of knowledge in science than Wallace and far more than Darwin's one sea voyage while he was studying to be a clergyman and didn't even know what the word "science" meant.

Despite Lamarck's break with the Bible and started the thinking about evolution founded on natural history, he was discredited, because anything that they could not understand was deemed supernatural (deistic, metaphysical).

However, Lamarckism had even bigger problems finding a scientific basis. What internal system stimulated new directed traits to appear? The radical geneticists said there were fixed genes *and no other,* which Mendel had only described, fixed *traits* which were inherited in a set ratio. Therefore, at the time Lamarckism could be easily argued down on this point alone—until now, when it can be proven.

When Kammerer came out in support of Lamarckism and inherited acquired characteristics, evolutionists sat up and listened at his uncompromising studies. He became the biologist to watch as he was making a name for himself by reporting clear-cut experiments of inherited acquired characteristic. This brought the rage of the neo-Darwinists and he was attacked with all the ammunition they had. Kammerer was an intellectual type not given to counterattacks and defending himself. He finally broke down after four years of constant onslaught of his insatiable accusers. He only knew how to gentlemanly try to explain his results to neo-Darwinists which were sight-blinded to any rational argument. After Kammerer was driven to suicide (for whatever reasons), and having seemingly discredited his work, the issue seems to have been settled in their favor. We can see why this antagonism to acquired characteristics was so vehement. Any opposition to the established theory was attacked for serious reasons given above. If the accepted theory was proven false, it could have serious consequences. Weismann whose claim to fame was cutting off mice tail for fifty generations would be disproven. He had no interest whatsoever in being convinced that his own work was a failure.

The sad ending of Lamarck is that he died in poverty and buried in an unmarked grave, despite being the father of evolution.

Chapter III

Natural Selection – Genetics and Random Mutations

Natural Selection

Then, a new materialistic theory of evolution by Alfred Russel Wallace, published in 1855 and sent to Darwin in 1858 from the jungles of Indonesia hoping wealthy Darwin's position in England's high society would help promote the theory. Darwin did promulgate the theory as his *On the Origin of Species* (1859). It was totally materialistic. Variations in traits follow a bell curve and happened by chance and if one or more of the more extreme trait variations fit better by chance in the same or new environment, they would thrive as a new species and the lesser fit would die off under the tautology—survival of the fittest. Darwin's credentials for supposedly having written this theory rested on a single voyage where, when starting it, admitted that he didn't even know what the work "science" meant. His education was for the clergy, not science and he intended to finish his clerical training after the voyage.

Darwin was on the voyage primarily to have a good time away from his father as an unpaid employee to keep the captain company at mealtimes who wanted someone of like high station in society. The last captain had committed suicide due to loneliness. English captains never associate with the crew or others on board. Darwin was not hired as a natural scientist, which he assumed for his own prestige. Being educated as a gentleman, he and the captain got on wonderfully together. And, as a companion to the captain gave him leeway to direct the natural specimens that they were expected to collect to his friends *before* going to the British Museum. He learned much from these friends *after* his return.

Alfred Russel Wallace came from a low middle class family and educated himself in the jungles of the Amazon and later the jungles of the Indonesian archipelagos while collecting specimens for wealthy collectors in England such as Darwin. It was in vogue for the wealthy to collect something. Wallace was extremely bright but had no status in society. His theory may have been considerably different if he had studied life on the open plains rather than being too close to daily struggles in the jungle to see the thousands of years of equilibrium that existed in both the plains (more obvious) and the jungles. Malthus had written his book on food supply and population, which negatively influenced both Wallace and Darwin (Huxley's "survival of the fittest" which Darwin appropriated and many other things said by others).

However, now materialists had ammunition to attack Lamarck. The battle started against Lamarck's theory that *nonmaterial* urges and strivings could affect change in living things. Natural selection was so materialistic, that it could run on a computer today. It became the ruling theory in England and America. Russia adopted Lamarck from the very start as a natural process of development of all living things. No one questioned urges or strivings as motivations.

Professor Theodosius Dobzhansky became famous for his genetic "proofs" of natural selection.

Prof. Dobzhansky felt he had proven natural selection. He grew bacteria on an agar plate where an antibiotic had been placed and where it spread out, it killed the bacteria. However, at the very periphery of the spread of the antibiotic and the concentration was minimal, some bacteria survived with a minimal dose. This "proved" that a fortuitous gene variation in these bacteria could tolerate the dose of the antibiotic and would then survive *any* amount of the antibiotic thereafter. He could not explain how this tolerance created such a complete defense that it would overcome all doses of the anti-biotic after that.

The glitch in this experiment is obvious. Bacteria have well-developed internal defenses and exposed to a minimal dosage of the antibiotic or any toxin could organize defenses that then tolerate any dose of the antibiotic or toxin in question (like cancer cells). He felt that bacteria had a genetic variation for each and every toxic chemical they might encounter that allowed them to survive from whatever the source, rather than having—over billions of years—developed an immunity system to defend against such chemicals. His proof would suppose that bacteria have an infinite number of genes for all future antibiotics and toxin that they might encounter! Where do they store all such genes with their limited plasmids?

His other proof is rather extreme, but seems valid. Daphnia (a water flea) lives within a certain temperature range, and suddenly lowering the temperature too much, kill most of the Daphnia, but those that survive must have had a genetic variation for lower temperatures ranges. Certainly, this must be valid if the change is abrupt. But how often does the environment have only a single change so abruptly rather than several at once putting impossible demands on any genome. However, it could occur in rare instances, but not as a general proof.

There are many problems with the theory, which were noted from the very start but ignored, but one in particular—how do you find variations for something entirely new when there are no variations to choose from? Where does a new trait start? They had no answer. It is a serious problem, but they had no solution and simply avoided discussion of it.

The second major problem dealt with the tautology "survival of the fittest," (Who are the fittest—obviously, those who survive. And who survives, but the fittest!). While this is almost ludicrous, it is actually an integral part of the Darwin theory.

The third major problem deals with probability. What is the probability of these few extreme traits ever coming together, at the right time when needed if the entire process is pure chance? Evolutionists were increasingly disillusioned with the theory. One prominent biologist, Ludwig von Bertalanfly came out and blatantly said that the theory was so vague with insufficient verifiably criteria that are usually applied to scientific theories that he felt should fall in the province of dogmas with the reigning theory of materialism, utilitarianism, and economics. It was also a theory based on chance and probabilities.

The first to see its weaknesses was Darwin himself. Darwin abandoned natural selection in preference to Lamarck's acquired characteristics which shifted the 6th edition of *On the Origin of Species* in that direction. He confided in a letter to his cousin Galton that each year he found himself more compelled to revert to the inheritance of acquired characteristics of Lamarck—because chance variations and natural selection were apparently insufficient to explain evolution. He had lambasted Lamarck's inherited acquired characteristics in his youth and now coming around to seeing its value.

Most Darwinists fail to read his 6th version of *On the Origin of Species* which shows this change in thinking. In general, evolutionists were abandoning his theory in droves during the latter part of the 1880s.

There are many other problems with the theory. The theory was almost buried and abandoned by materialistic academics until around 1890s when a new genetic theory incorporated genetics to support natural selection—Bateson's new neo-Darwinism. However, by 1924 Darwinists came to realize that Mendelian genetics did not explain the origin of species. Bateson died a bitter man feeling that it was a mistake to have wasted so much of his life on Mendelism. He equally detested inherited acquired characteristics and did the most to destroy Paul Kammerer.

Genetics

The newly discovered work of the Capuchin monk Gregor Mendel, in Moravia had established the fact that basic trait of plants and animals are discrete and when cross-fertilization with another plant of the same species would show up as a fixed ration of their traits, and now called the Mendelian ratio. It too became the basic dogma. Later, genetics added that mutations of traits were completely random which avoided any metaphysical (subjective) attribute.

The organism was nothing more than a pawn in the process. It is not a theory of evolution and has no answer for new genes or new anatomical parts or functions. Mendel only described (but brilliantly) these fixed traits and nothing else. Later, these fixed traits were found to be genes on the chromosomes. It is still the standard interpretation of how genetic traits are passed on to the next generation but does not explain evolution by this process. *Nobel* Laureate Dr. William Niermann at the National Institute on Mental Health (NIMH) in Washington discovered the sequence of DNA. It was a major step in understanding genes and codons of which they are composed, Richard Crick who took the advice of Rosalie Tallman and discovered the helical structure of DNA was to cement the Neo-Darwinists dogma that proteins could only be made from fixed chromosomes but not the reverse, only to be refuted by viral infiltration of DNA.

In any event, the field of genetics became an established fact. Those that delved into Mendelian genetics created a materialist interpretation of his work far beyond any implications of Mendel's observations and used it to enlarge the materialistic theory of evolution. Bateson, a founder of neo-Darwinism had basically given up on finding genetics as a basis of evolution.

Random Genetic Mutations

These material geneticists came up with an idea (early in the 20th century but primarily noted after 1940s) to explain evolution through genetics. Genes mutated were produced randomly but at a regular rate (like a clock). The accumulation would provide numerous variations and natural selection *would fortuitously select* the appropriate mutations and create a new evolutionary species. However, this is pure conjecture, but believable except for the probability of such combinations coming together in any reasonable way or time span—the so-call over-looked "details" which destroy these theories. Sir Fred Hoyle and others felt that this thesis and natural selection were completely ridiculous for this reason.

One detail, not overlooked but totally ignored, is the possible effects that can occur to alter the resultant rate of mutations and their preservation. Many defect mutations go nowhere and die out. Radiation (solar) increases mutations, speeding up the so-called genetic clock that they rely on for timing past events by counting the number of variations since the previous specimens (if any are found or just blindly calculating it). And what happens to these variations when disease wipes out most of the species and most of these random traits are lost? The random mutationists know all about these complications, but simply ignore them. Worse, archeological evidence does not agree with conclusions based on the genetic clock (understandably). They continue to write articles given dates when such and such event must have happened based on this theoretical genetic clock.

While genes might mutate regularly, there is the problem of storing these variations over millions of years until they are needed. The answer has been that the "gene pool" is spread over a large number of individuals which begs the previous questions of disease and availability.

Others said that these traits would dilute themselves out of existence in a few generations with normal reproduction in a large species, being only a low percentage of the trait. Geneticists have ignored the criticism.

However, the materialists had a "convincing" theory according to the way they reported it, if you didn't look at the improbabilities that underlie their assumptions. Random mutations became *the pillar* of support that keeps natural selection alive. The society imbued with materialism accepts them, but are being attacked easily by the creationists in court.

Chapter IV

Recent Genetics

Epigenetics

The science of *epigenetics* stunned the genetic community and they did their best to squelch it for about 15 years. It finally became established and is now a major field of research. Epigenetics proved that factors beside environment can affect gene modifications. These factors turned out to be—of all things—internal psychic (nonmaterial urges and strivings) and serious chronic bodily ailments or stress!

Epigenetics was first reported years ago in a lay scientific journal mentioning that acquired characteristics of Lamarck was the right answer after all and was ignored. How is the public to be informed with this code of silence toward any opposing theories?

The genetic changes can be passed on to future generations. What was so terrible about epigenetics? It was about to destroy one of the two pillars of modern evolutionary theory—random mutation as a factor in evolution. It gave support to Lamarck's theory which was right after all.

Inherited acquired characteristics could be explained but not in any materialistic theory, but by *awareness, urges,* and *strivings*—down to the first cells in evolution—just as we evolved by being aware. By resolving urges and strivings—a totally *nonmaterial* characteristic of living organisms ("sentient interieur" as Lamarck described it; and not "intention" as the English prejudicially translated it to discredit Lamarck) could create change and new species.

The materialists are still numb about these findings. What could be worse than finding *nonmaterialistic* aspects of life (like the nonmaterial mind directing the body) when your entire belief system denies any such finding could exist? It had to be a deistic concept, but it isn't. Descartes was dead wrong! The nonmaterial mind, the senses, emotions, and the body are inseparable. Just as the nonmaterial mind directs the material muscles to move the body, epigenetics is the nonmaterial process that influences the material chromosomes directly for evolutionary needs. This leaves the second pillar of modern evolutionary theory—Darwin where he was in the late 1880s—sinking in quicksand.

Epigenetics was discovered by renegade geneticists. Now evolution could be explained by acquired characteristics resulting from awareness, urges, and strivings in the organism itself. Epigenetics has shown that *various* phenomena can affect gene modifications. Both physical *and psychological* factors (nonmaterial urges and strivings) *can modify gene production*. Only the end results in the chromosomes at present can be seen but the mechanisms of these epigenetic pathways have yet to be explored. The Holistic theory of evolution has postulated that these pathways are also *nonmaterial* and influenced by the entire awareness of the organism on the chromosomes. New species *and* new classes of organisms can be created by the epigenetic process in less time than natural selection demands. (See the evolution of lungs in bony finned fish to amphibians below).

As seen in paleontological findings, evolution occurs in *spurts*, not slow gradual changes as Darwin or Lamarck assumed. Several genes are modified in the direction of a given change—not one at a time. Where Darwin would have giraffes fortuitously having a variation of a trait for tallness and therefore could eat higher up on trees and become the fittest. It now appears substantiated that smaller animals that *see* food higher up and continually *strive* to obtain it, by increasing in spurts in evolution to obtain it via epigenetic pathways (supported by paleontological artifacts of the various spurts to arrive at their present height).

Note: The belief has been that gene mutations move as a very *slow steady* pace like clockwork which supported natural selection with new genetic variations to select. Paleontology proved that this is not the case at all, and move in *spurts* where an entire trait or part will appear all at one time and plateau for thousands, even millions of years if the spurt has achieved its objective and then allowing the *entire* organism to adjust to its new condition before making another spurt. Epigenetics continues to work constantly but subconsciously without the organism's awareness. The immediate needs take priority over evolutionary prospects that are put on a back burner. Urges and strivings motivate epigenetics as necessities, but influence genetic effects—subconsciously.

Here we have to divide genetics into two paths, the *immediate,* and the other *ongoing.* In the immediate, the biochemical pathways have been well worked out. From sensory receptors (stimulus) a messenger RNA (m-RNA) is sent to the chromosomes. There, a gene is opened up and a negative copy is made onto a transfer RNA (t-RNA) to the endoplasmic reticular where the molecule needed is to be made and then used. This process takes care of all immediate biochemical and material needs of the organism on a millisecond to millisecond basis. It is the main material production system of the organism and has been completely organized, since eukaryotes, which were the first to have a complete set of chromosomes.

The second pathway is the epigenetic pathway, which is poorly understood and was only found in the latter half of the past century, except its effect on the chromosomes and changes in the organism. The epigenetic process can add a methyl group to certain genes turning them "on" thereby up-regulating the amount of some molecules such as enzymes as amylases in saliva of species on high carbohydrate diets, or an acetyl group to turn a gene "off," thereby down-regulating the production of proteases (enzymes to clean teeth in the saliva of carnivores). The process is hereditary as long as the diet is consistent, but can revert back again if the diet changes profoundly. Changes are often noted in hindsight and how fast or slow they develop, is hard to evaluate. It appears to take hundreds to thousands of years for a major change.

The strange part of the epigenetic process is how it directly affects the chromosomes to modify, block, alter, add, or create new genes when that is a necessity (e.g., the gene or genes that created the thenar nerve and associated muscle in australopithecines, the lungs in amphibians, and other incipient forms or processes).

It is not motivated by immediate needs but by the overall subconscious chronic awareness of the organism responding to chronic urges, striving, and needs of the entire organism on the basis of survival. At times it is much faster when survival is at stake. Because it is nonmaterial it does not depend on the same processes as the standard material production process above. Its effects only show up over time in spurts.

Epigenetics (and quantum mechanics to be discussed below) have destroyed the concept of pure materialism. It would take a few hundred pages to explain and one can find it in *The Holistic Theory of Evolution* by Yuriy Kalinnikov.

However, if the reader can accept these newer sciences as valid, then we can discuss the studies that prove acquired characteristics are real and apply to all life.

The epigenetic process depends on the nonmaterial aspects of an organism. Kalinnikov's holistic theory of evolution give the fundamental force of evolution to epigenetic factors as hunger, danger and any other need for survival—to add or change the organism in some way; physically, mentally or socially. The process works by taking a totality of these factors that include the entire awareness of the organism and all its faculties and functions. As Baron George Cuvier stated, any change in evolution changes *everything* in the organism, no matter how imperceptible the changes are. The *entire* organism is involved no matter how small or large the change might be. This totality of the organism awareness is the crucial factor and the most difficult part to understand. However, it is little different from a nonmaterial urge which can move the entire organism.

The latest genetic proof of inherent acquired characteristics will be described at the end of the book.

Chapter V

Energy – Matter – Mind

Quantum Mechanics (physics)

This brings up the science that completely undermined materialism and scientism. This was the discovery of quanta by Max Plank in 1900 and expanded by Niels Bohr and Werner Heisenberg in the 1920s which found a totally nonmaterial third aspect of matter and energy called *mind.* This third aspect causes all the unpredictable and unbelievable observations found inherent in quantum physics.

It seemed so strange, that even Einstein called it "spooly science." Now it is considered the most reliable theory in physics.

What seems predictable in the everyday macrocosm was totally unpredictable in the microcosm of quanta.

Quantum mechanics is a radically new physics, and is the physics of the twenty-first century. It cannot be taken lightly despite its difficult concepts. It is no longer the simple classical physics of Einstein where everything moved by fixed laws. Einstein once said that to understand the laws of physics was to understand the mind of God. He was so wrong.

Electrons travel around the nucleus in pairs, called entangled pairs. If one spins in one direction, the other always spins in the opposite direction. If they are separated by great distances and the spin on one changes, the other *instantly* changes it spin as well. The old standard laws of physics cannot explain this phenomenon. The speed of light is far too slow to affect this instantaneous change. It requires a new concept to be introduced into physics, such as the concept of a continuum where everything in the continuum reacts on everything else in the continuum instantaneously—and, every *aspect* of the universe exists in every*thing* in the universe but in various forms (such as quantum mind of matter and energy which evolves to awareness in organisms and consciousness in higher mammals).

Unpredictability

Also, another problem arose with standard physics. Take an electron rotating around a nucleus. According to Einstein and other physicists of the old classical scientism school of physics, the electron was supposed to rotate in strictly circular energy orbits. When actual experiments were carried out, quantum scientists found that the electron path around the nucleus was totally *unpredictable* and revolved close or far away in vertical, horizontal or any angle in between—in short, its path created a fog around the nucleus. This *unpredictability* was called "mind-like"—not unlike a moth around a flame which has an unpredictable mind just as we do as well—and everything else in the universe. Therefore our nonmaterial aspects evolved from this unpredictability aspect that quantum physics call *mind* in the rest of the universe. Kalinnikov sees it as "awareness," the most basic aspect of any concept of *mind*. If all this sounds far out—science fiction, or unreal, it has been well proven! Unpredictability is inherent in all things because it is a basic quality of quanta. It is the basis of the *mind* and *creativity*. One has to accept it, or fade like scientism. If you find it difficult to accept, you are not alone. Einstein could not accept it either, and spent the last forty years of his life trying to disprove it and failed.

Along with other quantum phenomena, Roger Penrose, Henry J. Stapp and others evolved this concept into the human mind and the mechanics of thinking based on quantum wave effects, while David Bohm took it to its very essence at the quantum level as well as to the human mind. He described *mind* as basic to quanta as matter and energy. He included *mind*, matter, and energy as one indivisible unit of quanta—as one totality. However, to explore quantum *mind* of quantum physics (mechanics) and then jump from this quantum level to the mind of living organisms is no small task. It was Kalinnikov[1] who made this transition by realizing that a discontinuity had occurred that changed simple biochemical recognition of food and noxious substances by surface transducers (sensory receptors) of the growing prebiotic complex to subjective recognition of knowing that has meaning and significance to the proto-biotic complex and then throughout evolution (as well as evolving a "self"). As size and complexity evolve, atom become molecules and *mind* show additional facets as primitive awareness in this complex. From this primitive beginning this universal *mind* becomes the simple awareness of living organisms. The discontinuity is complete when the change from objectivity of mind in the universe changes to subjectivity in living organisms takes place.

---

[1] The Anatomy of Art and Stimulation – A Neurophysiological Exploration of Stimulation through the Evolution of Art, by Yuriy Kalinnikov, copyright 1990 and 2013.

Several new sciences have grown up in the past hundred years and their results are very revealing of contradictory findings regarding nonmaterial properties of energy and matter in particular. The lay scientific press or even professional journals do not often address them possibly because the concepts are difficult to explain. Of late, they have made even bigger cracks in the walls defending scientism, which is being totally undermined by this new basic understanding. If epigenetics was not bad enough, quantum physics was to upset the entire material world with more fundamental *nonmaterial* aspects of the *entire* physical universe that even Einstein could not accept, but proved to be correct. It has led to a completely new interpretation of the physical world and explains the nonmaterial basis of life, awareness, and consciousness—the strangest of all phenomena, a factor that could only be explained as *mind* integral to the material universe. It has shaken the materialist to the core. In addition, while they still try to interpret life in mechanical terms, they are quietly fading into oblivion, but are still trying to explain life mechanically, just as Einstein was still trying to find a "theory of everything" mechanically for the rest of his life but failed because he couldn't accept the new science. Since there are still believers, they are not without a fight, like socialism, which is an incomplete faith based on materialism.

The world of the ultra microcosm is so weird that it will take years before the public can digest it. The world of the quanta disobeys most of the laws of the old physical sciences (now called "classical" physics) as opposed to the new reigning quantum physics. Not that the laws of physics are no longer valid, but their exclusive material basis has been totally undermined with new interpretations and findings of the basis of nonmaterial factors in the universe that gave a basis for the unpredictability of energy and matter, but also found *mind* (awareness) in living beings.

Kalinnikov found another aspect of awareness in neurophysiology that had been ignored throughout evolution because of the belief in materialistic natural selection and random mutations where the organism itself doesn't count—a pawn of material forces of random mutations and natural selection—happenstance.

Every proto-biotic complex had to have sensors to recognize food and repulse toxic substances based on electron densities of the evolving sensors. In evolution they evolved into sensory neuron which has a sensory transducer (sensory receptors with enzymatic properties) that takes physical phenomena from the environment and changes them into *internal* stimulation for the organism's awareness to interpret for action. The environment does not stimulate the organisms. Rather, the organism *selects out* of the environment what phenomena it needs and changes it into internal stimulation. For example, there are myriads of electromagnetic radiations that have no effect on the organism whatsoever. It is our sensory receptors (transducers) that select out only those phenomena that help us evaluate and navigate the environment to find food, avoid danger, and help us survive—although we use these sensors today for a myriad of additional functions. Because of these sensory transducers we can see, hear, and feel the environment about us—in short, we become *aware*! The most amazing fact about these sensory transducers is that they evolved from the very first organisms to be created and existed in the millions or billions of proto-biotic complexes that were to become living organisms—making them, and all organisms in evolution *aware* because they *all* have sensory transducers (more and varied as evolution progresses) for the same reasons—to recognize food, to protect ourselves, and navigate in the environment making us aware of what we are doing—so too with every organism alive.

This brings us back to inherited acquired characteristic armed with new sciences that no longer support the old materialism or materialistic evolutionary theories.

Chapter VI

Material Biological Sciences versus Nonmaterialism

The Battle lines

It appears that too much was at stake in this battle of materialistic theories and the nonmaterialistic theory of Lamarck and the newer holistic theory by Kalinnikov that support acquired characteristics. No one likes to have their beliefs undermined by new concepts no matter how attractive they may be, particularly changing their philosophy and weltanschauung or relationship to society (for the same reasons religion found scientism intolerable but had to live with it). Materialists are still trying to destroy religious institutions today as well as Lamarck.

To think that this is merely an academic or scholarly debate over evolution is to miss the point. A famous example is the old classical science of physics, which Einstein championed, versus the new science of quantum physics of Niels Bohr and Werner Heisenberg. Einstein failed to disprove it. The new generations of young physicists went for the new quantum physics backed by grants, positions and support and the old guard quietly died off and the new physicists took their place.

In 1926, when Kammerer reported his findings, there were no concrete proofs for any evolutionary theory but assumptions, reasoning, observation (often limited) and conclusions based on a theory of how evolution is suppose to work.

The nineteenth century had no better concrete theory than natural selection, and lacking sound proof for their assumptions, they could ignore equally weak criticism. Sir John Frederick Wilhelm Herschel blasted Darwin's theory as a "law of higgledy-piggledy," because without genetics it was just that. At the time Darwin was devastated as Sir Herschel was his inspiration to become a scientist. Opposing theories were often criticized based on side issues, unsubstantiated conjectures, sweeping negative generalizations, personal accusations, and misinterpretations of data and experiments, accusations of sloppy experimental data or even accusations of falsified data (opinion was enough to start rumors). Kammerer was attacked with all the above ammunition by the materialist evolutionists. Even Kammerer's death was publicized as an admission of guilt. After discrediting his work, the issue seems to have been settled in favor of the failing materialistic theories of the times—that is, until the random mutation theory revived natural selection a third time.

If you look at the proof against acquired characteristics, it relies primarily on a single study. Weismann chopped off the tails of mice for fifty generations (1882-1892) and the fifty-first generation developed normally after that. It was considered "proof-positive" and repeated in many texts ad nauseum after that. If someone were to chop off fifty generations of human hands would the 51st generation be born with no hands? In hindsight, this study seems almost ludicrous. If our (or mice) chromosomes are left alone and only physical parts are chopped away which are needed, why should the chromosomes change when the animal needed that part to survive? There might be increased wound healing or increased anxiety on seeing someone in a white lab coat with a bloody hatchet in his hands, but nothing else.

The theories of materialists and Lamarck are not different by just some semantic interpretation but profound. They are as different as Genesis and modern evolution. Ever since Lamarck's theory, all theories are based on a progression of life from the simple one celled organism evolving upward to multicellular organisms on to more and more complex structures and capacities with increased sensors to define the environment and greater flexibility of movement to find food and avoid danger and protect themselves. If one will note, the materialists insist that there is no direction (metaphysical) but do ascribe to evolution which shows direction. From this point, of the difference in how these changes occur become great. One has to know what is behind the basis of each to understand the battle. However, first, we want to review what is already known by this time.

There have been many valid studies proving acquired characteristic that are ignored in print or discussion so that public is unaware of the true nature of acquired characteristics. More and more studies and new sciences are showing that Lamarck's conviction that urges and striving for food or protect themselves from danger *do affect* the following generations and was completely valid. And, more and more studies are showing that natural selection and random mutations are only based on documentaries that asked their own questions and give their own answers that prove little. Do any studies come to the public's attention as objective reporting on inherited acquired characteristic? Rarely. They come only through the editing of materialists and omitted by journals who also profess a material doctrine. Who said: "This knowledge must *die*?" They were not chattering facetiously. People tend to accept only what they already have been led to believe despite contrary factors.

Weismann a researcher famous for his one study of chopping off the tails of mice proved to *all* that acquired characteristics did not occur. The *press and scientific community* broadcasted the results. It repeated in most *textbooks* every generation and continues as "proof" that negated acquired characteristics. Moreover, any study that proved acquired characteristic had to be disproven or ignored. They used any situation to attack the theory, as Kammerer was to find out.

They have gone so far as having "sight-blindness" which is being coined from the biblical expression "thou seeth yet seeth not". In some cases of proof, they simply cannot see it. It is somewhat reminiscent of the doctor telling the patient that she is way overweight and need to use a diet which he described in detail which she agrees to, but before leaving asks: "Do I take this diet before or after my meals?"

All early rebellions tend to be simplistic in their views (Liberty and Fraternity, All men are created equal, Life Liberty and the Pursuit of Happiness, Each according to their Needs, etc.) and ignore facts that do not agree with their over-simplistic beliefs. Later when cooler heads prevail, the dogma shows its flaws. Never has this been more obvious than when probability was applied to random mutations.

On the other hand, defenses are ever present. The most frequently used defense, when results are above criticism, is to completely ignore them as if the evidence does not exist.

With the rise of Nazi Germany, they claimed that eugenics would create a "master race." However, cooler minds realized that genetic mutations could always create abnormal genes and new medical syndromes so that a "master race" could never be achieved.

However, the battle-line had been drawn between materialistic theories of evolution (natural selection and genetic random mutations) and the nonmaterialistic theory of inherited acquired characteristics (whether it be Lamarckian or the newer Kalinnikov's Holistic theory). And the battle was all out war. The stakes were huge—

grants, budgets, government subsidies, private funding, professorships, prestige, Nobel prizes, medals, gold stars—you name it.

Where the battle had been essentially won and seemed to be no longer contested, Kammerer appeared to shake up the entire scientific community, especially of biology and beyond. His work was convincing and gathering support of other scientists. His book gathered together long obscure excellent research that proved inherited acquired characteristics, physically, and mentally.

The materialists were making their last ditch stand, just as religions did against the onslaught of materialism. All major thinkers adhered to the materialistic theories, but some were listening...

Despite propaganda for their materialistic causes, many hangers-on remain in the science community, which are losing ground today among serious researchers. Even against religionists, they are losing court battle after court battle, which allows Genesis to be included in science courses with threadbare theories of natural selection and genetic random mutation as alternate explanations to explain evolution. Lawyers know all the flaws in natural selection and random mutation theories better than the material evolutionists. The lawyers have no blinders on. The materialists still dominate the news media and TV programs but more information that is more objective is getting through via the internet and self-publishing. (Money is no object versus the unsupported opposition.) They ask their own questions and give their own answers and in their magazines have beautiful photography and drawings all aimed to support the "in" opinions of the times.

Many biologists tended to support acquired characteristics (e.g., Redford 1923) but dropped this portion of their work when opposition became intense to these claims. It became an economic, financial, and governmental issue when these institutions adopted scientism as the basis of financial support for research and promotions. To live in this atmosphere, it was prudent to avoid this issue. With this controlling viewpoint, many accepted negation of inherited acquired characterization regardless of how unsubstantiated the conjectures were. The "unwavering" impersonal laws of physics ruled everything by definition.

Kammerer stated that there were numerous experiments which essentially dealt with inherited acquired characteristics, but just as today are "deplorably lacking in courage" to call it what it is.

There are such amusing names as: "cumulative after-effect" (Alverdes 1921), "oscillating mutations" (Cuénot), "transgressive oecologisms" (Lang 1909), "enduring modifications" (Jollos), adaptive evolution and possibly others. It is no different from today with fear of attacks from the materialistic natural selection and random mutationists.

Even awareness in lower species is treated with extreme caution the same way, using such words as "salience" to avoid saying a fly could be aware or have attention despite having the similar sensory transducers to ours and rudiments of a nervous system working as does ours. Instead of being a part of the twenty-first century, they cower in fear of the powerful materialists and their socialist allies. Or, feel they will be accused of re-discovering the wheel but have discovered a profound basic proof of how evolution works.

Earlier, paleontologists (O. Abel, Cope Osborn among others) were positive toward inherited acquired characteristics and O. Abel in his book *Palaeobiology* stated outright that the inheritance of mutilations (misprint of mutations?) was indispensable to evolution.²

---

² Quoted from Kammerer's book The Inheritance of Acquired Characteristic, 1926. The theory of Random Mutation was not established at that time.

Chapter VII

Inherited Acquired Characteristics

The Experiments

Kammerer also included valid experiments from other researchers that were ignored but putting them together made for a stronger case. For example:

G.M. Allen noted that different geographical locations due to altitude, moisture, such as the hot dry sunny exposure as noted in the south and Western United States changed the inheritance of birds and mammals. Birds and mammals change to a lighter coloration, which is inherited if the progeny stay in these desert conditions and intense sunshine. This hold true for the Colorado potato beetle (Leptinotarsa decemlineata) changes color in hot dry sunny conditions (Towers).

Christian Schröder (1903) observed the moth Gracilaria stigmatella, used the tips of willow leaves as a cocoon by turning the tip of the leaf down over them and sealing it with exudates they excreted. When the tips of the leaf were cut off, some wrapped themselves in the sides of the leaf, which they brought together and sealed it around them. They passed this acquired characteristic to subsequent generations who when given a choice, continued using the sides of the leaf.

Christian Schröder (1903) also forced the larva of the European Willow-beetle (Phratora vitellinae) to feel on downy willow leaves which they bored into rather than eat from the surface as with smooth willow leaf and the offspring inherited this acquired characteristic and immigrated to the downy willow leaf as their new home.

Pictet (1905) force a number of caterpillars of the species Ocneria (Lymantria) to change plants that they normally fed on. The new food plant changed the coloration in the subsequent butterfly that proved to be hereditary even when eating their regular plant food.

Kammerer reported several studies of force-feeding animals alcohol (beer) (Germany, Austria—where else?) which showed acquired characteristic in their offspring, but the one report of two dogs by Kahrhel where two dogs were accustomed to drinking beer had a litter of four puppies who insisted on drinking the mother's beer and refused to drink water.

Degenerative and Atrophic Changes

Because one of the criticisms against his studies was that, the acquisition of hereditary characteristics could mean regression (atavistic reappearance of ancient genes or just possibly progression) which can go either way and that he had not ruled out regression. He quoted several studies showing degeneration (atrophy) as well as progression that do occur.

Kapterew did a series of breeding experiments with *Daphnia pulex,* a little crustacean deprived of light. This caused their eye to lose its regular shape and become ragged around the edges. Bigger or smaller pigment matter separated and distributed themselves all over the body, eventually to vanish entirely by absorption. In the beginning, it started out as an incidental finding, but after fifteen months this condition spread to the entire specimens and it had become hereditary; for quite young Daphnia of only four or five days displayed almost entirely discolored, and therefore destroyed eyes.

There were studies in humans related to atrophic organs. Plate in 1913 showed inherited acquired characteristics in the reduction from previous size. Widersheim counted ninety of such remnants, once well developed and well functioning, such as the appendix, the crescent-shaped fold of the inner corner of the eye (former nictitating membrane of lower species). The coccyx, are the rudiments of a tail and muscles that wiggle the ears are used in many animals to turn in the direction of sound.

It appears that when there is a lack of need and stimulation, epigenetics no longer supports an organ or function, and it atrophies. While the association of the mind-brains and awareness are fairly straightforward, the new epigenetic pathways are a total enigma (but not supernatural). This aspect of epigenetics needs further intense investigation.

As an aside that proves how epigenetic works:

When fish tasted the high calorie grasses when land was inundated by spring flooding, they wanted more. To obtain it, they could only gulp a bubble of air in their throat, which absorbed a small amount of the oxygen that allowed those with bony fins to crawl out and eat some before needing to return to the water. Over time, and repeating this process from generation to generation, the throat mucosa enlarged around the bubble of air to form a tube and closed off the distal end. Eventually it sprouted saccules off the main tube so that more air could be taken in and formed a lung—*entirely separate* from the gill system. It was an incipient process with no random traits to come together. It was entirely new and created amphibians by the epigenetic process—because they had the urge and the incentive. Several genes evolved together to form this lung in a directional way (not randomly). Whether these pre-amphibian fish gulped in a bubble of air *intentionally*, one cannot say, but we do know that human divers gulp in extra air "instinctively" before diving to stay under longer to find oysters and pearls—and continue to do it *intentionally*.

On the other hand, mammals that have returned to the sea thousands of years ago, never developed gills. There is no urge or way for them to conceive (become aware) of gills nor a way to initiate such a process and epigenetics has nothing (no urge or inner awareness) to initiate such a process. Epigenetics *helped* streamline their bodies from their urge to swim better and helped them navigate well in water as well as conserve oxygen for long periods, but no gills. Eventually, some incipient process may evolve in another direction (absorb oxygen through the skin or other ways), but at present there is no hint of a solution.

Most of Kammerer's own experiments involved forcing various amphibians to live and breed in radically different environments and look for inherited acquired characteristics that resulted.

He raised two type of salamanders one black (Salamandra maculosa) that lived in the Alps giving birth to two fully formed offspring. The second was a spotted salamandra maculosa that lived in the lowlands and gave birth to 10-15 larvae that like tadpoles metamorphed to adult salamanders. He then subjected them to reverse environmental conditions. The offsprings were raised under similar conditions. All showed reversal of their reproductive habits, which was inherited by their offspring.

In the next ten year period, he raised another group of salamandra maculosa on different colored soils (black or yellow). The back salamanders were raised on yellow soil and the yellow salamanders on back soil. Over the years the black salamanders developed larger and larger yellow spots, which was inherited by their offspring and the yellow salamanders raise in black soil became almost completely black which again was inherited by their offspring.

Kammerer in 1910 found that lizard (Lacerta sepa) when kept at high temperatures, their color changed to black that was passed on to their offspring despite living at normal temperatures.

Kammerer also bred tree-frogs (Hyla aborea) in a hot dry environment that led them to deposit eggs in water instead of dry land which became hereditary.

In still another study, he used cave dwelling newts Proteus, which are blind with rudimentary eye deep under the skin. In daylight, they only developed black pigment over the eyes (protection?). He was clever enough to try red light that does not induce pigment, and the eyes developed perfectly and the offspring inherited the ability to see.

One study was of particular interest in the sea-squirt (Ciona intestinalis). It has two siphons for inhalation and exhalation along with food particles. Kammerer cut off one of the siphons in over a hundred sea-squirts and the siphons grew back, but from the side (not the cut end). In most cases it appeared larger than a controlled group of over a hundred normal sea-squirts. This is understandable in that they need the organs to survive, just as mice needed their tails. The fact that the regeneration was not strictly limited in size can be accounted for by phylogenetic reasons of undeveloped controls that one finds in higher species (e.g., mammals). However this study was duplicated by others that seemed to have been done to refute Kammerer, by observing that the siphons grew back at the same size proving the *more important point* that chopping off parts of species does not alter their chromosomes or development and *disprove* Weismann's famous contentions that inherited characteristics do not occur, because chopping off tails of mice for fifty generations does not affect the chromosomes.

An important study that Kammerer reported was carried out by Ivan Pavlov of dog conditioning fame was on the inheritance of acquired mental characteristics in mice, which was reported in 1923 at the International Congress of Physiologists at Edinburgh where Schroder presented his paper on acquired characteristics in moths.

Pavlov conditioned a group of mice, which received a piece of cheese whenever he rang a bell until they had been completely condition to salivate on hearing the bell despite not getting cheese. It was positive associations between food and an environmental cue, which epigenetics would reinforce (see above). It took three-hundred attempts to completely condition the mice.

The second generation was given the exact same experiment but it took them only one hundred attempts to have them fully conditioned and salivated whenever hearing the bell. The third generation took only fifty attempts to completely condition them to salivate at the sound of the bell alone and the last generation—only five!

Despite being reported again in California before the Battle Creek Medical staff in July 1923 it was sight blinded. Another questionable study showed controls also improved, but less is known about the controls. It was attributed to "living in a science laboratory (improved memory and ability resulted in conditioning?)."

As far as Pavlov's experiment was concerned, it had absolutely no effect whatsoever, and was ignored. His dog conditioning is well known.

Hopefully, this book will awaken the reader and others who are listening to the results and achieve a more impartial view.

Chapter VIII

The Establishment versus Kammerer

The Attack

This brings us back to the reason for Kammerer's demise. Kammerer's work which was not only attacked on all sides and accused of fraud to discredit all other experiments he had completed, but also the attacks on the man himself came fast and furiously. The initial criticisms were reported in Kammerer's book.

Kammerer[3] reported his experiments, which clearly showed proof of acquired characteristics. Whether or not all of his results were valid is not the issue here, but the counter-reaction to his findings. Anyone reading his book[4] could see he felt stumped forever trying to prove anything to these sight-blindness dogmatists who absolutely refused to see any basis for acquired characteristics despite his taking pains to address all criticism

---

[3] The entire incident was thoroughly investigated by Arthur Koestler in the book The Case of the Midwife Toad.

[4] The Inheritance of Acquired Characteristics by Dr. Paul Kammerer.

leveled. He was apparently a gentlemanly type of professor and never lost his dignity (at his own expense).

While he reported his results *in detail*, the counterattacks from Weismann and H. E. Ziegler, prominent figures of the day, rejected Kammerer's findings based on conjectured sweeping statements that his experiments showed no results of acquired characteristic at all. They insisted that there were only showing atavistic traits of more ancient species, even though *it was passed on as an inherited acquired characteristic from the parents* to offspring regardless of where the traits came from. The criticisms were often just diversions from the basic issue.

Weismann was none other than the Weismann of chopping-off-tails-of-mice fame. He had no choice but to sight-blind himself along with Ziegler. This was their only defense in the face of being disproven, and seeing their life's work and reputations go—up in smoke.

Kammerer summarized the criticism of the neo-Darwinists, which showed that their criticisms negated any possibility of ever proving acquired characteristics, with their *conjectures* and *unproven* suppositions: All objections were raised in such a form as to make it impossible to answer the question altogether. Opponents of acquired characteristics were convinced that acquired characteristics were undoubtedly the reappearance of archaic recessive genes, but with no proof of such a conjecture (Weismann and Zeigler) and dodged or ignored any experiments that they could not refute (Pavlov). Nonetheless, his studies had *proved* that such traits were passed on to subsequent generations. Whenever a response was given the same conjectures were their responses: 1) The germ plasm can always be reached directly and no need for the environmental influence. 2) No new characteristic need to be really new, but may constitute an atavistic trait. 3) Each and every new characteristic may be affected or intensified by selection as far as they resulted from experiments even without being of great duration (Darwin's slow evolution be damned). The blind newt Proteus experiment was negated by the fact that the animal is transparent and therefore the light could have affected the germ plasm directly. Proteus forebear undoubtedly had sight and this ancient gene was probably still in the germ plasm and could reappear without exposing them to red light. It was not light itself but *natural selection* (they said when all else failed) restored the eye (in matter of weeks!). Appearance of newly acquired characteristics meant nothing because it was only the internal *faculty to move* an organism one way or another independent of outward traits or the environment (early random mutations). 4) Physical traits meant nothing. (Bauer) The environment need not affect physical change by affecting the body (soma), but could directly affect the chromosomes (germ plasm) to produce change. This belief became the main argument against acquired characteristics (with no theory as to how this was supposed to occur). Actually all effects affect the chromosomes, especially if of long duration. 5) There were also the over interpretative Mendelian geneticists who insisted on the theory of exact inheritance (W. Johannsen) as well as the theory of mutations (De Vries and his disciples) who waltzed away whenever having to answer questions of the evolution of new phyla, classes, families, species or any incipient new appendage or function. The materialistic Mendelists attacked it on the basis of needing only a fixed set of genes for all times. That is simply not true. Genes can become (lungs), change, and grow (gills) with time (uncommon though it may be). In addition, it was criticized by those that had formulated this radical materialistic theory around Mendel's work and said it violated Mendelian inheritance by introducing new genes which up to that time, the numbers of genes was an inviolate fixed rule of genetics. This bunch soon faded from history

In Kammerer's book *The Inheritance of Acquired Characteristics,* he had summed up his work and that of others that had long been obscure.

On reading the book (now again in print), Kammerer comes across as a very mature gentleman, a sincere person who never reduced himself to angry replies or to name calling, and the book makes no radical or outlandish statements either for his work or against his critics. In fact, he seems rather restrained despite their attacks and tries to perform studies that answer their criticism in a serious way. One comes away with respect for the man, but wonder if he was too restrained (too timid or too gentlemanly?) Later, in defending himself against personal attacks of fraud, he did not attack but instead was taken aback, surprised, and bewildered by such accusations. All of his anger seems to have been turned inward at a terrible cost. Many individuals had access to his lab and specimens which could be used in his defense, but felt he could not for lack of concrete evidence.

His most famous experiment began his downfall. The materialists who attacked in full force used it deviously.

The experiment was to force the midwife toad to live under a very dry hot environment to see the effect on mating which occurred on land and the male carried the eggs as midwife until ready to hatch and then placed them in water. In the hot dry environment, they were forced to stay in the water where the eggs were free to move about. Mating also changed to mating in water all of which was inherited for six generations by their offsprings until the group died out but had proved the objective of the experiment.

He also *incidentally* noted that in water the male midwife toad developed calluses on the little finger with small spikes for griping the female in water called nuptial pads. This *incidental* finding was published and the fireworks began.

The most vicious attack was started and perpetuated against him on t*echnical details* and *not* on the results of the experimental findings that went by the wayside and ignored, but on a technical detail of an illustration. None other than Bateson, the bitter neo-Darwinists, who hated acquired characteristics and possibly growing fame of Kammerer even more, started it.

The issue was a photograph of the foot of a male midwife toad that had developed a nuptial pad on a finger to hold on to the female in water.

Bateson, now an authority on midwife toad led the attack. He insisted that one picture of Kammerer's toad experiment had been doctored (opinion). Also, it was not a midwife toad but another species (opinion). Despite Kammerer's confirmed explanations, Bateson continually repeated his assertions in the journal, *Nature*. Rumors started to spread about a "manipulated specimen and others joined in the attack. The news media found it sensational and gave it full coverage." Kammerer took the slides and sections to the National Historical Society in Cambridge and had several prominent biologists verify the authenticity of the photographs and specimens. Bateson refused to attend and continued the attack for years.

Bateson appeared to have to been a bitter man and while losing faith in neo-Darwin aimed all his hate against inherited acquired characteristics. He was the main attacker of Dr Kammerer with all sorts of innuendos and side-tract issues and diversions that had no real meaning since they did not touch the basis of the experiments that showed acquired characteristics. He enlisted Dr. E. D. Boulenger, a Belgium who was director of the reptile department of the British Museum who originally was favorable to Kammerer but was easily swayed with innuendos from Bateson and joined in to question (and later to defame) Dr Kammerer with incorrect investigations and comments in print that Kammerer refuted each time, because the accusations were totally unfounded. Much of this collusion with Bateson was carried out in private over dinners and unrecorded, except loudly reporting supposed failures and inaccuracies of Dr Kammerer's study of the midwife toad in *Nature*.

While the British Historical Society and many experts testified to the authenticity of the photographs and specimens, it did not stop Bateson but Boulenger dropped out of the controversy, but the damage had already been done and had its toll on Kammerer. The specimen in question rested in the Institute where the work had been done years before and understandably deteriorated. Many had access to doctor it, but proof had already been established. Bateson passed away, but the issue was far from over.

A critic from the New York Museum of Natural History, a hotbed of reactionary thought, by the name of G. K. Noble curator of the reptile department and ardent anti-Lamarckist, went all the way to Vienna to examine the specimen. The director and Kammerer gave him free access not suspecting the specimen had been doctored shortly before this examination was to take place. After examining the evidence of the male midwife toad, *publically accused* Kammerer of falsifying the specimen by injecting the feet with black India ink long after the specimens laid in the laboratory. Kammerer and the director were stunned by the accusation and insisted his specimens were valid and not doctored. They were unaware of any problem. Dr Kammerer was taken aback by the finding of apparent sabotage of his specimen. Regardless of his reaction and protest, the damage had been done, and he was disgraced by these pronouncements because of all the negative publicity it received. No forensics—nothing but personal opinions would decide! Anything, to discredit his statement that midwife male toad had evolved nuptial pads (black) to grasp the female better in mating in water, even though Kammerer was not using this observation to support his convictions of acquired characteristics. Kammerer wrote that it *looked* as if he had perpetrated the sabotage of his toads and could not prove otherwise (trying to see the situation objectively as others might see it), but fervently denied any responsibility. In a letter he wrote that he suspected who did it but couldn't prove it. Despite *disclaimers*, and *conformation* by third parties later after his death of his innocence, a reexamination of the photograph clearly showed the small spikes that are part of the nuptial pad. His career was destroyed and his book faded into obscurity and Dr. Kammerer was devastated. From a world figure, they had reduced him to a laughing fraud. He obviously could not express his anger and turned it inward.

What is most interesting is why at this late date was Noble so interested in the specimen, which *had to have been* doctored a few days in advance of his visit? Why had he made a special trip to Vienna just to examine an old specimen, already proven authentic by the British Historical society at Cambridge? Later, experiments proved that the toad's finger had to have been injected at most, only three days before Noble's visit. Otherwise, the India ink would have dispersed through the area and the media. Was Noble in on something for his own notoriety or benefit? Noble is primarily noted today for this incident. Anyone could have gotten into Kammerer's lab while he was no longer a faculty member. Why was he so outspoken about the finding instead of just say that it had been injected with India ink without accusing anyone? Certainly, Bateson could have had something to do with this trip but there is no evidence at this late date, and much of what Bateson did was far from above board or public. Were Bateson and Noble friends, or in communication?

Yet all of the counter-conjectures of the opposition against Kammerer were never questioned. Noble and others seemed too eager to silence him, apparently for their own personal convictions, and destroying another prominent investigator in the process. Did Noble care what he was doing to another person? Even Kammerer's apparent "suicide" was quickly interpreted as proof of guilt; but murder seems possible although he did intend to do away with himself with final letters denying any sabotage of the specimens. More likely, a botched assisted suicide was the result of the gun in the wrong hand rather than falling to the ground on the left side. Someone else had to have been present to put the gun in his right hand.[5] After

---

[5] The entire incident was thoroughly investigated by Arthur-Koestler- in the book *The Case of the Midwife Toad*.

that," it appeared to have silenced his work and that of others for good—at least for years! Nonetheless, his work did open the debate in acquired characteristics. Other researchers had come to the same positive conclusions on acquired characteristics and the subject would not go away, but of no consequence until this century, when detailed genetic studies were carried out to prove unequivocally the correctness of inherited acquired characteristics.

It was finally proved in the 21st century when detailed genetic studies confirmed that acquired characteristics are a reality—not randomly but with more rapid direction for the organism's needs (hints of this work was already appearing in the 20th century). The public has been so brainwashed into believing these pronouncements of natural selectionists as fact, but with no counter proof. The public tends to go along with the "authorities."

Now it is time to expose the issue and allow the public to read the proof versus conjectures for themselves. Nonetheless, before that, the critics had their hour to strut upon the stage, but then, are heard of no more. Bateson, Weismann, Ziegler and Noble are grains of sand, in the sands of time that record great men.

In the early 19th century no one could really prove or disprove acquired characteristic without detailed genetic studies to confirm or rule out the success of random mutations or Lamarck's. Now we have the proof.

The reigning natural selection theory stated that marked random variations in traits would fit better in some (future?) environment, better than others and survive as the fittest. Thus, evolution would evolve slowly *by chance* over millions, or needed billions of years, according to the acerbic Sir Fred Hoyle.

While this controversy was boiling, World War One had ravaged the Austro-Hungarian Empire and Vienna the capitol lay bankrupt. Kammerer was left penniless and his lover refused to go to Russia, which greatly depressed him. He ended going on a speaking tour to raise money. It was hugely successful. He was gaining popularity, which was already substantial and drew large crowds whenever he spoke. He continued his speaking tour in America. He was not only famous but had a large scientific following. The materialists had to get rid of him or their own goose was being cooked. He is best known for keeping inherited acquired characteristics alive despite the relentless, intense campaign to discredit them.

Chapter IX

Discussion

Inherited Acquired Characteristics

From the above, one has good reason to feel that evolution is about inheriting acquired characteristics and not some change trait finding a new environment or a goddess of natural selection "gathering up all the good and rejecting all the bad" as Darwin metaphorically stated (in place of St Peter?). And, as Sir Fred Hoyle pointed out—for random mutations to fortuitously form a DNA molecule would take probability about $10^{40,000}$ years! With the direction, we see in living forms and the progression of awareness to consciousness—to believe it is all just chance or luck, really stretches the imagination and completely negates rational evolution. Instead, natural selection and random mutations would give evolution no choice but to go randomly forwards, backwards, sideways or nowhere.

However, one must add that unfortunately laboratory studies on amphibians or insects only bring out the proof that acquired characteristics *are* inherited which is all they were meant to do. Having established this fact we arrive at the first step in understanding evolution as a rational process. The question is—how do we acquire new traits that evolve? Here we run into heated arguments from scientists and new research that rejects solution solely based on materialism

We will continue this issue under the discussion below.

We can understand acquired characteristic by noting our entire being is the result of acquired characteristics from our distant and recent evolutionary past. Just about everything we are has been inherited as acquired characteristics from previous generations, our biochemistry, anatomy, physiology and our orientation in dealing with life. With rare exceptions, everything we are has been acquired—our lifestyle, our scientific advances our world as we find it. Cities, roads, stores and things we buy have been acquired from previous generations with the exception of the latest electronic gear and fashions. Even our place in history, the conflicts and advantages have been acquired from past generations. We see first of all it is both physical, mental, and social.

Socially, we can see from above how each generation acquired traits, benefits and disadvantages from previous generations, not in some random fashion, but by direct influences and consequences. Our inner conscience is a product of the first hominid societies that lived on the open plains. The largest mind-brain homo sapien were the Neanderthals who introduced us to art on cave walls, caring, and compassion for the elderly as seen in their graves. Every generation has passed something on to another as the Neanderthals passed their culture to our species—modern Homo sapiens.[6]

On a more personal level, our family traits, our ethnic type, and identity is acquired from our parents and our place in the family determines a lot of our character. There is our immediate environment as schooling, friends, and way of life as it affects us as an acquired characteristic. We would not be the same person in another country during the same time.

Contrary to idealists, we are neither equal nor the same as other people all around the world with

---

[6] For those who feel the above facts are bizarre, recent evidence proves the statements are correct. Not only that, but the Neanderthals wore clothes, not bear skins, and never clubbed their wives or dragged them home as cartoonists and modern egotists want to believe, or that modern Homo sapiens were the beginning and end all of civilization. While we may be the "end all" of civilization, we are hardly the beginning of civilization, or speech. Speech is also an inherited acquired characteristic probably from the australopithecines.

only different dress like naively portrayed by UNICEF holiday cards trying to perceive us, dancing together around in a circle. We have acquired convictions from our educations, family and friends. We have acquired different beliefs and standards, especially our ethics that often clash with others.

Let us see acquired characteristics as an incipient trait or structure where no trait existed before. There are basically three types: 1) Superficial traits that come and go as desired or needed, 2) semi-permanent traits that exists for hundreds or thousands of years but when the environment changes can revert back or disappear, and the 3) are permanent genetic changes that last for millions of years.

Temporary acquired characteristics have been commonly noted in stylistics. A tribe adopts common attire which is socially inherited for the life of the tribes as a rigid social trait for the all important need of identity. It is part of social evolution.

Sun worshippers who sunbathe to excess develop temporary defenses by increased melanocytes as long as they stay in the summer sun, but production returns to normal during the winter when they no longer sunbathe. These temporary acquired characteristics are physiologically and genetically driven and based on biochemical defenses from previous generations.

The semi-permanent inherited acquired characteristic changes are a subconscious epigenetic phenomenon. Major changes in a long term diet, such as from primarily protein to an essentially carbohydrate diet, demands different salivary enzymes. Such changes need more amylase enzymes in the saliva in primarily carbohydrate diets and needs less proteases to keep the teeth clean, but if there is a major change back to meat over time, the salivary enzymes can change back again without our awareness.

Permanent inherited *incipient* or other acquired characteristics also involve the subconscious epigenetic system to produce new genes for the new functioning parts. For the natural selectionists and random mutationists, they have nothing to say and ignore the issue (and they have become experts in waltzing).

The famous case is an incipient gene (or genes) for the thenar nerve and muscle of the thumb in australopithecines which allowed them to grasp things (e.g., spears) for the first time, not found in primates and allowed them to form aggressive-defensive societies on the open plain of Africa starting the trek toward civilization. For this, the subconscious epigenetic processes played the primary role. Another is the lung that was discussed above.

According to Lamarck, acquired characteristics were the result of urges and striving to utilize new food sources out of reach (like the giraffe) which affect the body that he thought, slowly changed continually as they are passed on strengthen parts to the next generation. This is not entirely correct. Paleontologists find that the continual striving over periods of time *do* cause evolution but in *spurts*. Then perhaps thousands year elapse to adjust to all the subtle changes before another spurt is made (Baron Georges Cuvier).

The theory propagated by Darwin is a mechanical, materialistic, deterministic, nonpurposive concept where the giraffes happened to be an extreme natural variation of a trait for tallness and happen to be able to take advantage of food on high trees and became a species of the fittest to survive. (Today this sounds like so much –after the fact reasoning.) It certainly does not match up with paleontological observations.

But what is happening in the meantime? Having established that we *do* inherit acquired characteristics, we want to know, how it came about. What is the nature of evolution and this strange process of epigenetics? This process was not even thought of during Kammerer's lifetime and no one would have believed it. It appears almost completely nonmaterial like *mind*—and probably is.

Chapter X

The Deeper Meaning

Implications

Here we come back to the deeper and more difficult challenges between the material and the nonmaterial world of which we are composed. Many can no longer see the solution simplistically as a deity plopping everything down and easily explaining everything by projecting all issue back to this deity.

We know we are not only material things but *living* beings with feeling, emotions and thoughts which materialists try to ignore, but they have to be considered as well—and in all living beings, plants *and* animals. The more we investigate, the truer this fact becomes.

After years of indoctrination into thinking materialistically, and to reorient ourselves to a holistic concept of life is not easy, but the change is inevitable. We still have our material side but now we have to consider in a rational sense our nonmaterial side, which is integral to the material. What is the strangest part of all is the fact that *the nonmaterial part moulds and runs the material side,* not the reverse. Because it sounds deistic, the materialists thought they could eliminate it in their over simplified theory to rebel against supernatural religious convictions, even though the nonmaterial aspects of life are just as sound as the material, but to understand them and their relationship is difficult. However, that is no excuse to dismiss them just because there are no easy answers. In the past century scientists on the forefront of this new understanding of the universe have been struggling to understand these strange unpredictable aspects in the universe themselves and some gave up trying (Einstein) while other persevered and are still struggling but continue to peel back inner layers of the microcosm to find—whatever.

These issues are now *our* issues to deal with and try to understand. As we read above, we see that these enigmatic influences on genes and chromosomes produce these acquired characteristics and an even stranger force that quantum physicists can only describe as *mind* that is integral to all matter and energy and together create our material and nonmaterial universe and all life.

But now, we have no deity, no high priest to whom we can turn to for an explanation. Perhaps Lamarck was leaning partly on his deistic background when he felt that evolution was leading to perfection—the image of a deity? However, we no longer use that concept for answers unless one has always believed in a deity. We have to integrate ourselves into these new understandings with what we already know. The challenge of mind, matter, and energy is ours alone. It must have been like the first australopithecines standing out on the open plains for the first time and seeing the vast distances, mountains, and sky as they had never seen them before living under the jungle canopy. They too had to integrate this enigma—alone.

To try and see how we are created out of these new discoveries is far more difficult and profound than the simplistic material theories of the past and our relationship to each other and the universe with which we are now intrinsically one is no small undertaking.

Earlier, others found simplistic materialistic answers that they found satisfactory by negating anything they could not understand is not good enough today. If one chooses to open one's eyes, we see new horizons as vast as the physical one the australopithecines saw in the past. We know there is more, and the new sciences force us to re-examine every theory objectively and to accept that we are not at the end of the search nor are answers concrete, but as new discoveries have reached deeper into our understanding of the nature of matter and energy (and now *mind*), we can be certain that there are more layers to the onion to be peeled away until we come to the core (if ever) and can find the ultimate theories to live with. To believe in natural selection and random mutations and accept simplistic explanations based on post hoc reasoning is the easy way out, and not worthy of the challenges that the new science of epigenetics that prove inherited acquired characteristic are real but only found out in the past century. If anyone is to integrate these sciences with the present, it is only we ourselves that have the task.

As others have stood up to be counted, facing rebuke and ostracism, we admire those who stand up for their faith and beliefs whether they are main stream or not, or agree with our convictions. Are they right by a majority of one?

Chapter XI

Kalinnikov's Holistic Theory

Holistics

We tend to think of ability to focus as being the most important part of our being and ignore all the workings of the body and subconscious as one complete unit. Our conscious awareness is supported by every part of us and cannot be separated. We are not the same person with a single attribute missing though it may appear slight. Our awareness is affected in some way. It is more obvious when we lose an arm or leg, or an eye, but our awareness is modified. It may be more theoretical than in a practical sense when we lose an appendix in surgery, but we shouldn't minimize the effect nonetheless for a very important reason. It is our *total* awareness (consciousness and subconscious) that influences the epigenetic system. It takes in all aspects of the organisms before it acts to influence the chromosomal changes.

Because of this concept, one has to change one's thinking of parts and systems and only of the entire organism and all biochemical, physiological, and psychological systems acting together, whether it is a one celled prokaryote or all the cells in the human body acting as one—because they *are* one totality, working inseparably together. When the organism (or human) has a need or desire, the entirety of the organism has the need or desire, and the epigenetics process is the coordination of this awareness (*mind*) in the direction of fulfilling the need or desire, by a constant sensory and material exchange with the environment as the conscious mind controls the everyday actions and needs of the body.

But it does not end there. We are a product of the earth and its environment and gravity. We are inseparable and have been molded by these factors, just as the earth is influenced by the sun and solar system and our galaxy and finally the universe of which we are an integral part. The laws that govern the macrocosm and the microcosm also govern us and are reflected in our unpredictability and creativity.

Mind

This totality is the *mind* of the one celled organism and the mind of the entire mind-brain-body of humans—not the chromosomes in the nucleus, the physical brain, or any other proposed centers, but the entire being as one—is the mind. It is only *focused* through the mind-brain.

For this theory, we need the *mind* discovered by quantum physicists to evolve our minds. The *mind* is no longer metaphysically separated from the body as Descartes naively proclaimed. It is the process of acquired characteristics, incipient or otherwise. The biochemistry has not been elucidated and may not exist, but the motivation for it has. As with the mind, there is no process that intervenes between the mind and the neurons—they are one.

It sounds metaphysical until one understands the nature of awareness. This has been explained in Kalinnikov's books *The Holistic Theory of Evolution* and *The Anatomy of Art and Stimulation,* (a much more involved discussion of the neurophysiological basis of art and stimulation), and both require a brief description of quantum physics (mechanics) to explain the basis of awareness from the more basic quantum physics and the concept of *mind*.

On the other hand, one of the strongest proofs of inherited acquired characteristic was carried out in bacteriophage in raising the temperature, but the researchers were apparently reluctant to come out and call it what it was (in the face of possible intense materialists attempts to discredit their results or ignore any reference to it, or for fear that they had rediscovered the wheel?). [7] But, by any other name ("adaptive evolution"), it was still a Lamarckian inherited acquired characteristic. The research was well planned, tedious, and laborious laboratory work, and brilliantly carried out. It will be covered extensively below.

Epigenetics

Kalinnikov's interpretations it based on the nature of epigenetics—and awareness, which is based on the mind, evolved from basic quanta.

The biochemical process (if any) is only what one sees as the end result in the changed chromosome, which have been described above but not the motivation for *directed* change. And, the process has to be valid for the first cells as for humans.

---

[7] Attacking the researcher, his technique or insisting on unfounded suppositions are not uncommon tactics of the materialists.

Genetics – Introns and Exons

Kalinnikov's theory is the third and the most controversial theory of evolution yet to evolve.[8] It proposes that when an entirely new *internal* need or urge develops, a subconscious epigenetic process stimulates introns and exons to create entirely new modifications of genes or even entirely new genes for entirely new organs (lungs in amphibians)—entirely new traits for the first time such as the thenar nerve and muscle in the thumb of australopithecines in order to grasp things—using the thumb as we do today. Its evolution had to be far more rapid and fill a need.

The materialists continue to believed it occurred by their regular but random mutation of random genes being disadvantageous, neutral or positive and natural selection would fortuitously amass the positive ones in this process, but demanding millions of years of chance combination of codons to form a gene; that has no purpose but *might* fit somewhere—not evolving to solve an organism's need, but might—fortuitously.

---

[8] *The Holistic Theory of Evolution*, the Evolution of Mind and Body by Yuriy Kalinnikov. Copyrighted 1990, 2007, and 2013 revised.

On the other hand, Kalinnikov proposes that epigenetic processes create *several* modifications of genes together to perfect a nerve or organ such as the lungs in amphibians in a specific direction and fairly rapidly *from internal need and desires.*

In recent years, major discoveries have been made in all areas of research, which should change the way we look at evolution, and not through the eyes of outdated theories. Because both natural selection and the random mutation theories try overly hard to be purely "classical materialistic science" (science prior to quantum physics), classical physicists feel that everything must be "objectively" proven, and anything that hints of metaphysics or the unexplainable has to be eliminated, whether it fits or not. One needed new equipment, more understanding, and observations of the way genes work before this could be done, but now investigations all over the world are bearing fruit. But even the above *explanation* of Kalinnikov's holistic theory is over simplistic. We can no longer think in terms of simplistic theories with finite effects as described by random mutations by merely producing mutations and letting natural selection collect them–period (end of explanation and the theory).

The holistic theory goes far deeper to the very essentials of life, merging them; that entails the very essence of the materials that are being merged, which turns out to be the universe itself. Mind, matter, and energy are not simply factors that evolved in organisms (into awareness and consciousness), but have their beginning in the very essence of the universe. The ramifications are profound and integrate us as one with the universe and its fundamental nature of which we are an aspect, not something separate. We change with time but never leave the universe only change to a new or lesser, differentiated form which Hindus and Buddhists symbolize with cremation. Where materialists had us as insignificant grains of sand on the ocean floor we are always integrated with everything.

The effects of epigenetics in moving evolution give us a direction—toward even more profound issues to be dealt with. We are not only abstractly integrated with the universe, but factually connected to this force that is evolving this universe and moving us as well, in ways we still do not understand, but does not appear to be material but purely nonmaterial aspect of mater and energy as our mind move the body and is also part of the totality. When we drive a car we become part of the machine and act as a whole. Without us, the car is meaningless.

We have no clear idea of where we are going, yet we are striving to get there in some possibly new evolutionary direction—to be cogs in a greater wheel with no solution to ultimate controls? To accept a holistic theory is to expand our orientation to a far wider understanding of evolution than the simplistic theories of the past offered with "and that how it is—just so stories." Do we take on the challenge or just turn on the TV?

We can finally return to acquired characteristics and see that it is the organisms *themselves* either consciously as in tribal stylistic identities of social evolution, or subconscious in humans for the sun-bathers protection, as well as to protect the bacteriophage from an elevated temperature environment, or the giraffe's desire to eat at higher levels or *intentionally(?)* as pre-amphibian fish wanting to stay longer on land to eat grass as pearl divers *intentionally* want to stay longer under water. Epigenetics represents the *mind* that influences the acquisitions and direction of all organisms. By being subconsciously aware of the needs and desires, it creates the acquired characteristics by the entire organism affecting the chromosomes directly, just as our nonmaterial minds can directly move the entire physical body.

These findings support inherited acquired characteristics by nonmaterial processes as epigenetics. In recent genetic research at the very level of genes, we see that natural selection and random mutations are nowhere to be seen in the process of the organism's need to change.

Chapter XII

Final Proof of Acquired Characteristics

Being convinced that once geneticists were able to analyze every variant gene that arises for some specific purpose, they would find that the number of positive *directed mutations* that move toward solving some problem at hand would far outweigh random ones.

In fact, this has already been done using viruses called bacteriophages, where the entire genome is known and each genetic change can be examined in detail.

After subjecting strains of a bacteriophage to a new controlled environment of thirty-seven degrees Celsius, the bacteriophage changed fairly rapidly to adjust to the new temperature, and changes in genetic make-up were analyzed. Several different genotypes were studied so that one could rule out uniqueness. After adjusting to the new temperature, the rapid changes in the temperature regulating genes plateaued. Fitness by number of offspring remained the same as prior to the test, showing that fitness was maintained. If this isn't directed genetics, an inherited acquired characteristics then it must be supernatural, because the random rate of other genes did not change proportionately. The rate of random mutations could not keep pace with the directed temperature regulating gene changes.[9] What produced the direction if not some inner awareness in the bacteriophage of this new environment and the need for inner defenses to take over. What additional proof does one need to accept directed evolution from within the organism itself, and, at the level of a single-stranded DNA (ssDNA) virus working with all its parts as one? This Nobel-quality investigation was done by Rokyta, Abdo, and Wichman in America. **These geneticists have knocked down random mutations the one pillar of twentieth century evolution that kept natural selection alive.** It was neither natural selection nor random mutations that changed the bacteriophage. **The bacteriophage themselves changed to adapt to the environment, at some primitive level of awareness, which Kalinnikov claims is the basis of acquired characteristics via epigenetics.** The investigators did not venture the underlying basis for the genetic change other than how the experiment was conducted. Kalinnikov sees this

---

[9] See Rokyta, Abdo, and Wichman in *Journal of Molecular Evolution* 69: 229–239.

work in the context of his holistic theory and finds it in complete agreement and supports the theory. Changes and direction come directly from within all organisms because they are aware and adapt to changes as we do! If it is freezing, we put on winter clothes—we don't wait for random genes and natural selection to grow fur on our bodies and random mutations and natural selection had at least 285,000 years to make a start, or a million years when Homo erectus migrated north and east.

However, if new nucleic acids or codons were added to the genes for this change to occur, these must have come from some neighboring source, if any were associated with the active genes around this single-stranded DNA. In plasmids, there are no apparent introns or histones, and needed mutations such as those possibly seen in bacteriophages may come from their intake of such compounds or produced internally as needed. Perhaps, internal rearrangements of genes without outside molecules were possible.

What is more, the paths that various bacteriophages took were **unpredictable**. They did not mutate the same as other bacteriophages, but ended up with the same adjustment. There were clearly directed mutations, but these were executed unpredictably, as one would expect from *unpredictability* built into all life. To try and see this process as pure physical biochemistry without some degree of motivated awareness is inconceivable. In addition, if awareness is not present here, how do we incorporate it in the future, when we experience awareness in our own conscious state?

These studies indirectly support *awareness, direction,* and *unpredictability* that is basic for Kalinnikov's holistic theory of evolution.

Unpredictability is inherent in everything in the universe and responsible for creating new configuration in genes and later new imaginary thoughts in humans. The change noted in this study was that while they all arrived at the same end point, they took various unpredictable biochemical paths to get there—not randomly, but unpredictably, in that it could not have been predicted beforehand what particular path each would take. This point is important when explaining aspects of epigenetics and the effects of quantum physics in the macrocosm.

With these observations in hand, it can be demonstrated that inherited acquired characteristics and the theory of holistic evolution are valid where direction and awareness of all organisms can be included.

This directed acquired characteristic changes has finally been proven in bacteriophage, an encapsulated single stranded DNA molecule where every gene and codon is known thereby solving the problem of how acquired characteristics are biochemically created for the first time. Not until this work on bacteriophage had been completed could anyone prove the validity of previous arguments.

Even here, there was apparently reservations in calling it Lamarckian inherited acquired characteristic but rather designated it as "adaptive" evolution which clearly relates to acquired characteristics due to a direct interplay with the environment and "sentiments interieur" to create physical genetic changes (Lamarckism). However Kalinnikov adds that in very primitive organisms, the epigenetic process involves primarily the *biochemical mind* which is the totality of the organism.

Despite all of the positive experiments showing acquired characteristics, that show that total materialism is false, and only superficial observations and conclusions to prove random mutations, materialists continue to insist that there is no direction and only randomness and fortuitous natural selection (interpreted it as chance) prevailed in evolution. It was the new "in" belief during the 19th century, and the overall convictions of the 20th century in science. The great minds of the periods advocated, as well as the news media and the public who also believed it. The other would-be-greats (e.g., Kammerer) were discredited with abstruse and conjectured opinion and completely ignored or scientifically assassinated. Kammerer was possibly driven to a stupidly assisted suicide. Positive results are buried while at the same time negative results are readily taken and trumpeted as fact. Again, there was no one (or grants) to repeat these experiments to see that they were reproducible. On the other hand, a highly improbable experiment was reported called "cold (atomic) fusion" in a small laboratory experiment. Various laboratories around the word rushed to repeat it (with no end of grants) only to end in failure but it fell under the banner of scientism.

We will see this overzealous approach proved to be its short-comings and cooler minds found out why acquired characteristics are essential for evolution.

Will this result in some famous person getting up and proclaiming the fact that scientism and random genetic mutation theories are dead? No, they will be swept under the rug as the last forty years of Einstein's failures were swept under the rug and all other dead theories before them. Some form of science will continue to evolve which at present is quantum mechanics or quantum wave physics as prophesized by David Bohm, and despite new studies by materialists being shot up like fireworks of new and wonderful things to come while quietly sweep their past mistakes—under the rug.

In conclusion, one has to come to the obvious realization that materialism as a concept is dead and that random mutations as a theory of evolution is dead, and natural selection can be an exception at times but without support has almost died in the past—including with Darwin himself. Giraffes are not simply an extreme trait that fit a higher food supply by chance, but evolved in spurts through urges and desire to eat what it could see but was not tall enough to obtain which spurred them on.

End

www.ingramcontent.com/pod-product-compliance
Lightning Source LLC
Chambersburg PA
CBHW051724170526
45167CB00002B/788